TALES FROM AMERICANA

Sasquatch

BY LAURA STICKNEY

Kids Core
An Imprint of Abdo Publishing
abdobooks.com

abdobooks.com

Published by Abdo Publishing, a division of ABDO, PO Box 398166, Minneapolis, Minnesota 55439.

Printed in the United States of America, North Mankato, Minnesota.
102023
012024

THIS BOOK CONTAINS
RECYCLED MATERIALS

Cover Photo: RichLegg/E+/Getty Images
Interior Photos: Shutterstock Images, 4–5, 10–11, 12 (left), 12 (middle left), 12 (middle right), 12 (right), 16, 28 (top), 29 (top); AP Images, 6; Braunger/ullstein bild/Getty Images, 8; Michael Wheatley/Alamy, 14, 28 (bottom); Victor Habbick Visions/Science Photo Library/Getty Images, 18–19; John Zada/Alamy, 20; Lloyd Young/AP Images, 21; John Bazemore/AP Images, 23, 29 (bottom); Michael Kappeler/dpa/picture alliance/Getty Images, 24; Bettmann/Getty Images, 26

Editor: Angela Lim
Series Designer: Katharine Hale

Library of Congress Control Number: 2023939655

Publisher's Cataloging-in-Publication Data

Names: Stickney, Laura, author.
Title: Sasquatch / by Laura Stickney
Description: Minneapolis, Minnesota: Abdo Publishing, 2024 | Series: Tales from Americana | Includes online resources and index.
Identifiers: ISBN 9781098292881 (lib. bdg.) | ISBN 9798384910824 (ebook)
Subjects: LCSH: Sasquatch--Juvenile literature. | Yeti--Juvenile literature. | Animals, Mythical--Juvenile literature. | Folklore--Juvenile literature. | Cryptozoology--Juvenile literature.
Classification: DDC 398.24--dc23

CONTENTS

With more than 450 sightings, California ranks as one of the top US states in Sasquatch sightings.

CHAPTER 1

Mysterious Footprints

It was a summer day in 1958. Jerry Crew was working for a logging company near Bluff Creek in northern California. As his team cleared trees, Crew noticed that there were huge footprints on the ground.

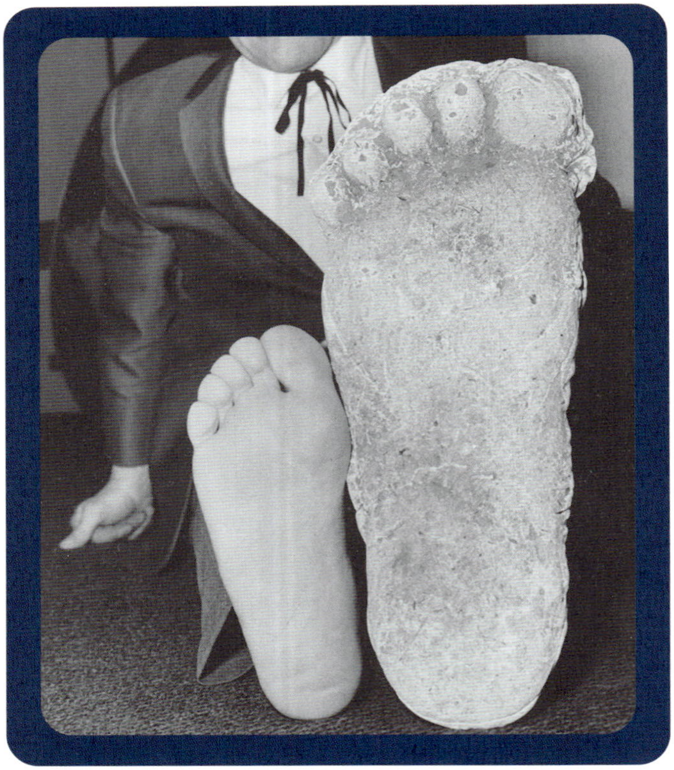

If Bigfoot wore shoes, it would wear a men's size 26.

They looked like human footprints. But they were much bigger.

Crew showed the footprints to other workers. Some of them had heard about similar prints found in the area. But who—or what—could be big enough to make the footprints? Was someone pulling a prank? Or was a mysterious creature lurking nearby?

Later, the workers found more footprints. One worker measured them. They were 16 inches (41 cm) long. That was much bigger than a human foot. The workers joked that a giant creature made the prints. They called it Bigfoot. A story about the footprints was published in a local newspaper. The news quickly spread across the country. Soon everyone was talking about Bigfoot.

Fake Feet

At first, many people believed the Bluff Creek footprints were real. But the prints were revealed to be a **hoax**. A worker named Ray Wallace had carved giant wooden feet. He used them to make the prints. He wanted to prank the other workers.

The Squamish people have a museum in Washington that celebrates their culture. They have stories about Sasquatch.

Sasquatch Stories

The Bluff Creek footprints made Bigfoot well known in the United States. People tell stories about a giant, hairy, humanlike creature. People call this creature Bigfoot or Sasquatch. Some believe there's just one Sasquatch. Others say there are many.

Tales of these creatures are not new. For centuries, Sasquatch have been part of American Indian legends. These creatures are **sacred**. Many American Indian nations in the Pacific Northwest and Canada have Sasquatch myths. This includes the Salish peoples. The name *Sasquatch* comes from the Salish word *se'sxac* or *sasq'ets.* This means "wild man" or "hairy man."

Today, Sasquatch still fascinates many people. Some try to find it and prove it exists. No proof has been found. But that doesn't stop people from searching for the creature.

Explore Online

Visit the website below. What new information did you learn about Sasquatch that wasn't in Chapter One?

Sasquatch

abdocorelibrary.com/sasquatch

Sasquatch is said to look humanlike and be covered in dark hair.

CHAPTER 2

The Legend of Sasquatch

Many people think Sasquatch is a **primate** that walks on two feet. According to **eyewitnesses**, Sasquatch is at least 6 feet (1.8 m) tall. Some reports are of a creature that is more than 10 feet (3 m) tall. Its feet are up to 24 inches (61 cm) long.

How Big Is Sasquatch?

According to most reports, Sasquatch stands about 10 feet (3 m) tall. This is much taller than the average man. It's also taller than a bear standing on its hind legs.

Sasquatch is covered in red or dark-brown hair. Many people say Sasquatch smells like rotten eggs or meat. Others describe the smell as being like a wet dog or a skunk.

Sasquatch is believed to live in forests in Canada and the northwestern United States. It eats plants and animals such as deer. The creature is shy and good at hiding. It walks silently.

People say Sasquatch makes noises, such as screams, whistles, and howls. Some people believe Sasquatch talk with each other. They knock rocks or branches together to communicate.

Sacred Creatures

Many American Indian nations believe in Sasquatch. Different nations have different beliefs and traditions about the creature. In some nations, Sasquatch is a **guardian**.

People from Coast Salish cultures tell stories of Sasquatch.

It protects the land and reminds humans to respect nature. The Sts'ailes people live in southwestern Canada. They believe Sasquatch, or Sasq'ets, is a shapeshifter. It can transform into different things. Sasq'ets moves between the physical world and the spirit world. Humans should respect it.

In some legends, Sasquatch can tell the future or become invisible. In other stories, Sasquatch is violent. It kidnaps and eats children. People told these stories to warn children not to wander into the forest at night.

Many versions of Sasquatch in movies and books do not represent American Indian cultures. In 1929, J. W. Burns wrote a newspaper story about Sasquatch. He talked to American Indian people about their encounters with Sasquatch.

Sasquatch around the World

In North America, people tell stories of Sasquatch. Other places have similar legends. In Asia, a snow giant is said to lurk in the Himalayas. It's called the Yeti. In Australia, people tell stories of the Yowie. Florida also has its own version of Sasquatch: the Skunk Ape.

Some versions of Sasquatch show the creature as only a scary monster.

But Burns retold their stories incorrectly. He did not talk about the creature's spiritual meaning or connection to the land. Instead, he made Sasquatch seem like a monster.

Burns did not represent American Indian beliefs. Many American Indians say this takes away from the meaning of Sasquatch. But Burns's story helped shape non-Native versions of Sasquatch that exist today. Sasquatch stories continue to grow and change.

Primary Source

Kelsey Charlie is a member of the Sts'ailes nation. He says Sasquatch is sacred to his people. Seeing Sasquatch is good luck:

> If you're able to see him, hear him, or see his footprints, there's some type of good [luck] that's going to come your way. . . . [Sasquatch] live off the land, they live on the land, they *are* the land.

Source: Lisa Kadane. "The True Origin of Sasquatch." *BBC Travel*, 22 July 2022, bbc.com. Accessed 12 June 2023.

What's the Big Idea?

Read this quote carefully. What is its main idea? Explain how the idea is supported by details.

The legend of Sasquatch interests many people.

CHAPTER 3

Searching for Sasquatch

Many people want to find Sasquatch. They collect **evidence** to prove the creature is real. **Plaster casts** are one type of evidence. People made casts of the Bluff Creek footprints.

People have tried to film or photograph Sasquatch.

Bigfoot hunters may take plaster casts of large footprints they find.

In 1967, Roger Patterson and Bob Gimlin filmed a short, blurry video of Sasquatch near Bluff Creek. Many people thought the video was fake. They said it was a person wearing a gorilla suit. But others believed it was real.

In 1976, a man sent hairs to the Federal Bureau of Investigation (FBI). He said they were Sasquatch hairs. The FBI tested them. They turned out to be deer hairs.

In 1998, an associate professor at Illinois State University studied hair samples to see if they belonged to Bigfoot.

Sasquatch hunters may join organizations. They go to meetings where believers share evidence. They also look into sightings.

Bigfoot or Bear?

People may mistake bears for Sasquatch. A study found that sightings of Sasquatch were more common in areas with many black bears. Black bears sometimes stand on their back legs. This may give them a similar appearance to Sasquatch.

One Sasquatch-hunting group is the Bigfoot Field Researchers Organization. It investigates Sasquatch sightings across the country. Most sightings occur in Washington, California, Florida, Ohio, and Illinois.

Many people believe evidence of Sasquatch is fake. No Sasquatch body, hairs, or bones have ever been found. This makes many people doubt that Sasquatch exists.

Pop Culture Icon

Sasquatch has been featured in many movies and TV shows. In some depictions, it's a monster. For example, Sasquatch appeared as a monster in a 1976 episode of the science-fiction show *The Six Million Dollar Man*. Some films

Sasquatch items are displayed at the Sasquatch Museum in Cherry Log, Georgia.

show a friendly Sasquatch. In the 1987 film *Harry and the Hendersons*, a gentle Sasquatch befriends a family.

Legend the Bigfoot was the official mascot of the 2022 Athletics World Cup, which was held in Eugene, Oregon.

Sasquatch has also appeared in children's movies. The 2019 film *The Missing Link* features a Sasquatch named Mr. Link. He teams up with an explorer to find his family.

Reality TV shows have also explored Sasquatch legends. The series *Finding Bigfoot* follows a team of Bigfoot hunters. They travel the United States to investigate sightings.

Many cities have Sasquatch-themed festivals. Marion, North Carolina, and Willow Creek, California, host yearly Bigfoot festivals. During these festivals, people participate in events such as Bigfoot-calling contests. Whoever makes the best Bigfoot noises wins. Some towns also have stores and streets named after Sasquatch. Several states have Bigfoot museums.

Roger Patterson said he saw Sasquatch in California in 1967. He took a video of the event.

Fact or Fiction?

People continue to debate whether Sasquatch exists. Most scientists say there isn't enough evidence to prove Sasquatch is real. There are no good photos of the creature. But believers say this is because Sasquatch is good at hiding.

Even without proof, many people believe Sasquatch is out there. They keep searching for it and sharing stories. The creature remains popular, leaving a big footprint in people's imaginations.

Further Evidence

Visit the website below. Does it give any new evidence to support Chapter Three?

Is There Any Proof That Bigfoot Is Real?

abdocorelibrary.com/sasquatch

Legendary Facts

Sasquatch is a giant, hairy, humanlike creature. It is also called Bigfoot.

Many American Indian nations have legends about Sasquatch.

Sasquatch lives in forests and is good at hiding from humans.

Many people want to find Sasquatch. They collect evidence to prove the creature exists. But no proof has been found yet.

Glossary

evidence
information that proves that something is real or true

eyewitnesses
people who see something happen with their own eyes

guardian
a protector

hoax
a prank or trick meant to fool people

plaster casts
copies of imprints, such as footprints, made by pouring a material called plaster into the prints

primate
a type of animal such as a human, monkey, or ape

sacred
holy or worthy of respect

Online Resources

To learn more about Sasquatch, visit our free resource websites below.

Core Library CONNECTION
FREE! COMMON CORE MULTIMEDIA RESOURCES

Visit **abdocorelibrary.com** or scan this QR code for free Common Core resources for teachers and students, including vetted activities, multimedia, and booklinks, for deeper subject comprehension.

Booklinks NONFICTION NETWORK
FREE! ONLINE NONFICTION RESOURCES

Visit **abdobooklinks.com** or scan this QR code for free additional online weblinks for further learning. These links are routinely monitored and updated to provide the most current information available.

Learn More

Gleason, Carrie. *Guide to Bigfoot.* Crabtree, 2023.

Krantz, Laura. *The Search for Sasquatch.* Abrams Books for Young Readers, 2022.

Index

About the Author

Laura Stickney is a writer, editor, and artist from the Twin Cities area in Minnesota. She is currently pursuing a master of fine arts degree in creative writing.